Contents

1 Chapter 1: Soil's Vital Role

Healthy soil is the foundation of thriving agriculture. It's not just dirt; it's a complex living ecosystem teeming with microorganisms, fungi, and other organisms vital for plant growth and overall ecosystem health. A healthy soil structure, rich in organic matter, creates ideal conditions for water retention, aeration, and nutrient cycling – essential for robust plant development and high yields. Ignoring this fundamental truth leads to decreased productivity, environmental damage, and ultimately, economic instability. Understanding this intricate web of life beneath our feet is the first step towards sustainable farming practices.

The soil's ability to provide nutrients to plants is a remarkable process, a continuous cycle of decomposition, nutrient uptake, and release. Organic matter, the decaying remains of plants and animals, plays a crucial role. Microorganisms break down this organic matter, releasing essential nutrients like nitrogen, phosphorus, and potassium, which plants absorb through their roots. This natural nutrient cycling reduces the need for synthetic fertilizers, lessening environmental impact and promoting healthier soils. Healthy soil acts as a reservoir, releasing nutrients slowly and steadily, ensuring a consistent supply for plants throughout their growing season. This is in stark contrast to synthetic fertilizers which provide a fast, intense burst of nutrients that can be inefficient, damage soil structure, and pollute water sources.

Consider the soil's texture. This affects everything from water infiltration to root penetration. Sandy soils drain quickly, potentially leading to nutrient leaching and drought susceptibility. Clay soils, conversely, can become compacted and waterlogged, hindering root growth and aeration. Ideal soil texture involves a balance of sand, silt, and clay, creating a crumbly structure that promotes both good drainage and water retention. The perfect texture allows for optimum root penetration which leads to better access to nutrients, and improved water holding capabilities, reducing the need for supplemental irrigation. This ideal balance promotes the growth of beneficial soil organisms that further improve soil health and fertility, ensuring robust plant growth and high yields.

Soil structure significantly impacts the health and productivity of the land. A well-structured soil has aggregates—small clumps of soil particles—bound together by organic matter and soil microorganisms. These aggregates create pore spaces crucial for water infiltration, aeration, and root penetration. Poor soil structure, often caused by intensive tillage, leads to compaction, reduced water infiltration, and impaired root growth. This can result in reduced yields, increased susceptibility to drought, and a decline in overall soil health. Investing in soil health, therefore, demands techniques that preserve and enhance soil structure, promoting long-term productivity and environmental stability.

Furthermore, the organic matter content is not just important for nutrient cycling and soil structure; it's also key to soil's water holding capacity. Organic matter acts like a sponge, absorbing and retaining water, making it available to plants during dry periods. High organic matter content helps reduce water stress, leading to improved crop yields and resilience against drought. This natural water retention capacity is critical in the face of increasingly unpredictable weather patterns and water scarcity, a growing concern across many agricultural regions. The enhanced water retention also diminishes the need for intensive and inefficient irrigation practices, conserving a vital resource.

It's crucial to realize that healthy soil isn't just about plant growth; it's essential for the entire ecosystem. The soil supports a vast array of organisms, forming a complex food web vital for ecosystem services. These organisms help decompose organic matter, regulate nutrient cycling, suppress plant diseases, and enhance overall soil health. Disrupting this delicate balance can lead to soil degradation, reduced biodiversity, and decreased agricultural productivity. Sustainable farming practices must prioritize the protection and enhancement of this intricate web of life within the soil. Ignoring this fundamental aspect undermines the very foundation of healthy agriculture and compromises long-term sustainability.

In summary, understanding soil's vital role is fundamental to successful and sustainable agriculture. Healthy soil is not merely a substrate for plant growth, but a dynamic ecosystem that underpins the entire agricultural system. By understanding its complex processes, structure, and the critical role of organic matter, we can adopt effective practices that enhance soil health, increase productivity, and build resilient agricultural systems capable of withstanding future challenges. This knowledge forms the cornerstone of all subsequent chapters, paving the way for a deeper exploration of sustainable agricultural practices.

1.1 Healthy Soil Basics

Healthy soil is more than just dirt; it's a living ecosystem teeming with microscopic organisms, a complex web of life essential for plant growth and overall agricultural success. Think of it as a vast, intricate city bustling with activity, where bacteria, fungi, and other microorganisms work tirelessly, breaking down organic matter, releasing nutrients, and creating a fertile environment. This vibrant community is the foundation upon which thriving crops are built. Ignoring this intricate biological community is like ignoring the infrastructure of a city; eventually, the system collapses.

Understanding soil texture is crucial. Sandy soils, for example, drain quickly, potentially leading to nutrient leaching and water scarcity, demanding different management techniques than clay soils, which can become waterlogged and compacted, hindering root development. Loamy soils, a happy medium of sand, silt, and clay, often provide the best balance of drainage, aeration, and water retention. Knowing your soil type allows you to tailor your practices to maximize its potential. Ignoring this crucial aspect of soil composition will lead to suboptimal results, no matter how effective your other farming practices may be.

Soil structure, the arrangement of soil particles into aggregates, directly impacts water infiltration, aeration, and root penetration. A well-structured soil boasts a crumbly texture, facilitating easy water movement and providing ample spaces for air and roots. Conversely, compacted soils impede water infiltration, leading to runoff and erosion, and restrict root growth, limiting nutrient uptake. Improving soil structure is often achieved through methods like no-till farming and the addition of organic matter. The benefits are considerable, leading to healthier plants and a more resilient ecosystem. Neglecting this aspect can lead to significant yield reductions and environmental degradation.

Organic matter is the lifeblood of healthy soil. Composed of decomposed plant and animal material, it serves as a crucial food source for soil organisms, enhancing their activity and boosting nutrient cycling. High organic matter content improves soil structure, increases water retention, and enhances nutrient availability. Adding organic matter, whether through composting, cover cropping, or the application of manure, is a fundamental step in building soil health. Failure to incorporate enough organic matter compromises the entire soil ecosystem. The consequences of ignoring this vital component are significant, affecting everything from nutrient availability to water management.

Soil pH, a measure of soil acidity or alkalinity, significantly impacts nutrient availability. Most plants thrive within a specific pH range; outside this range, nutrient uptake is hindered, potentially leading to deficiencies. Regular soil testing allows you to monitor your soil's pH and make necessary adjustments, ensuring optimal plant nutrition. Ignoring soil pH is akin to ignoring a critical diagnostic tool, leading to potential yield losses and plant stress. Understanding and managing soil pH is a crucial step in achieving optimal agricultural production and is something that many beginners often overlook, leading to avoidable setbacks.

Soil testing is not merely a one-time event; it's a continuous process. Regular testing allows you to monitor changes in your soil's health over time, helping you track the effectiveness of your soil management practices and adjust your strategies as needed. Testing reveals essential information such as nutrient levels, pH, and organic matter content. Armed with this information, you can make informed decisions about fertilization, irrigation, and other soil management techniques. This continual monitoring is an investment in long-term soil health and productivity.

Nutrient cycling is a continuous process where nutrients are exchanged between soil, plants, and organisms. It is a dynamic process that involves several steps, including mineralization, immobilization, and nutrient uptake. Understanding this complex interplay is crucial for maximizing nutrient availability to plants. It's a delicate balance and requires an understanding of nutrient interactions and the role of microorganisms in the process. Failure to understand nutrient cycling leads to inefficient fertilizer use and potentially nutrient deficiencies.

Maintaining soil health is not just an environmental responsibility;

it's a cornerstone of profitable and sustainable agriculture. It impacts every aspect of farming, from crop yields and water usage to long-term farm profitability. Investing in soil health is an investment in the future, ensuring healthy crops and a thriving agricultural sector. By understanding and implementing these basic principles, you lay the foundation for a successful and sustainable agricultural journey. A commitment to soil health yields dividends both ecologically and economically. The long-term benefits far outweigh the initial effort and investment.

1.2 Nutrient Cycling

The lifeblood of healthy soil, and by extension, thriving agriculture, is nutrient cycling. This intricate process involves the constant movement and transformation of essential nutrients, moving from organic matter to plants, then back to the soil, creating a continuous loop of life. Understanding this cycle is critical for boosting soil fertility naturally and reducing reliance on synthetic fertilizers, a key tenet of sustainable farming. Ignoring this natural process risks depleting soil nutrients, leading to lower yields and environmental damage.

Plants, the primary consumers in this system, absorb nutrients dissolved in soil water. Their roots act as conduits, drawing vital elements like nitrogen, phosphorus, and potassium – the "big three" – along with many micronutrients, from the soil. These elements then become the building blocks of plant tissue, fueling growth and reproduction. The quantity and efficiency of this uptake are directly related to the health of the soil itself, its microbial activity, and its water retention capacity.

After harvest, the remaining plant material – roots, stalks, leaves – becomes a valuable resource. This "crop residue" contributes significantly to the next stage of the cycle, decomposition. Through a complex process involving a diverse community of microorganisms – bacteria, fungi, earthworms – organic matter is broken down. This process releases nutrients back into the soil, making them available for future plant growth. The healthier the soil ecosystem, the more robust and efficient this decomposition process becomes, mimicking the forest floor's natural nutrient cycling, where fallen leaves nourish the soil year after year.

Several practices can enhance this natural cycle. Cover cropping, discussed in detail in Chapter 2, is particularly effective. Planting cover crops between main cash crops adds organic matter and improves soil structure, directly impacting the decomposition process. Their deep roots also improve soil aeration and water infiltration, further benefiting nutrient availability. The choice of cover crop depends greatly on your specific location and soil needs.

The addition of compost, a rich mixture of decaying organic matter, boosts nutrient cycling considerably. Compost acts as a slow-release fertilizer, providing nutrients gradually to plants, mimicking nature's own timetable for nutrient release. It also improves soil structure, creating a better environment for microbial activity and root growth. The proper composting method—a careful balance of carbon-rich and nitrogen-rich materials—is a crucial skill for any aspiring sustainable farmer.

Understanding the soil's microbial community is crucial to effective nutrient cycling. These microscopic organisms are responsible for the decomposition of organic matter and the release of essential nutrients. A healthy and diverse microbial community thrives in a well-structured soil with adequate organic matter. These microorganisms are the unseen workers driving the entire nutrient cycle; their health determines the soil's ability to provide nutrients.

In contrast, conventional agricultural practices often disrupt this natural process. Intensive tillage, for example, can damage soil structure, reducing microbial activity and hindering nutrient cycling. The frequent use of synthetic fertilizers, while providing a quick boost to nutrient levels, can also disrupt the delicate balance of the soil ecosystem. This can lead to nutrient imbalances, reduced microbial diversity, and soil degradation. Long-term use of these synthetic compounds also presents negative environmental impacts, which chapter 7 will cover in more detail.

The efficient recycling of nutrients in agriculture is not merely about soil fertility. It also reduces reliance on external inputs, contributing to cost savings for farmers. Reduced fertilizer costs translate directly to higher profits. Moreover, it fosters environmental stewardship by decreasing pollution from synthetic fertilizers. By harnessing the natural processes of the soil, you are reducing your carbon footprint, minimizing water pollution from runoff, and contributing to a healthier planet. The goal is not simply higher yields, but the creation of a sustainable and resilient agricultural system.

This careful balance ensures that nutrients are available for plant growth without harming the environment or depleting the soil's long-term fertility. Understanding these processes empowers farmers to make informed choices, leading to healthy, productive soils and, ultimately, a more sustainable food system. It's about understanding the interconnectedness of all the elements within the soil. A healthier ecosystem, a healthier planet. The focus on nutrient cycling is a cornerstone of creating resilient and sustainable agricultural practices for the future. Ultimately, it's about nurturing the soil and empowering it to support us and future generations.

2 Chapter 2: Regenerative Practices

No-till farming represents a radical shift from conventional plowing. Instead of turning the soil over, which disrupts soil structure, damages beneficial organisms, and leads to increased erosion, no-till methods leave crop residue on the surface. This acts as a protective mulch, reducing evaporation, suppressing weeds, and improving soil health over time. The benefits extend beyond immediate soil improvement; reduced fuel consumption and labor costs contribute to a more sustainable and economically viable operation. This approach requires specialized equipment, but the long-term gains in soil fertility and reduced inputs often outweigh the initial investment. Many farmers have reported significant increases in crop yields and reductions in erosion after transitioning to no-till practices. The learning curve involves understanding the nuances of seed placement and weed management, but numerous resources and support networks are available for aspiring no-till farmers.

Cover cropping is a powerful tool for improving soil health and fertility. By planting non-cash crops, such as legumes, grasses, or brassicas, between cash crop cycles, farmers can replenish soil nutrients, suppress weeds, prevent erosion, and improve soil structure. Legumes, particularly, have the added benefit of fixing nitrogen from the atmosphere, reducing the need for synthetic nitrogen fertilizers. The choice of cover crop depends on specific soil conditions, climate, and the following cash crop. Careful planning ensures the cover crop doesn't compete with the main crop, and instead enhances its growth. This technique can significantly improve water infiltration, which is crucial for drought resilience. Observing the growth of cover crops provides invaluable insight into soil health, revealing areas needing specific attention.

Crop rotation is a cornerstone of sustainable agriculture, offering a multitude of benefits. By alternating different crop species in a planned sequence, farmers can break pest and disease cycles, improve soil fertility, and reduce weed pressure. Different plants have varying nutrient requirements and root systems, leading to improved soil structure and nutrient cycling. For example, planting legumes one year helps replenish nitrogen in the soil, benefiting the following year's crop. This diversity also fosters a more resilient ecosystem, less susceptible to widespread pest infestations or disease outbreaks. While planning a rotation requires careful consideration of market demands and crop suitability, the long-term advantages in terms of soil health and yield stability make it a worthwhile practice for any farmer. Successful crop rotation hinges on understanding the nutritional needs of different crops and the impact their root systems have on the soil.

Integrating these three practices – no-till farming, cover cropping, and crop rotation – forms the foundation of regenerative agriculture. It's not just about individual techniques; it's about creating a synergistic system that works with nature, rather than against it. These methods, when implemented correctly, lead to healthier soils, higher yields, reduced reliance on synthetic inputs, and enhanced resilience to environmental challenges. The long-term vision necessitates patience and adaptability, as soil health improvements are gradual. However, the cumulative benefits, from enhanced water retention and reduced erosion to improved biodiversity and increased profitability, make regenerative practices a powerful investment in the future of agriculture. Adopting a regenerative approach requires a shift in mindset, from focusing on short-term gains to cultivating a long-term relationship with the land. This requires careful planning, accurate record-keeping, and often, collaboration with other farmers and agricultural experts. The journey can be challenging, but the rewards—a healthier planet and a more sustainable food system—are immeasurable.

2.1 No-Till Farming

No-till farming is a revolutionary approach that minimizes soil disturbance, drastically altering traditional agricultural practices. Instead of plowing or tilling the soil, seeds are planted directly into the residue from previous crops. This simple change has profound and far-reaching effects, creating a healthier, more resilient agricultural system. The benefits are numerous, significantly impacting both the environment and the farmer's bottom line.

Leaving crop residue on the soil surface offers several key advantages. First, it provides a natural mulch, shielding the soil from the harsh elements. This mulch helps regulate soil temperature, preventing extreme fluctuations that can damage sensitive roots and hinder growth. The protective layer also reduces water evaporation, conserving precious moisture, especially vital in drier climates. Furthermore, this residue acts as a buffer against erosion, preventing valuable topsoil from being washed or blown away by wind and rain. The preservation of topsoil is crucial, as it contains the majority of the soil's organic matter, essential for nutrient cycling and overall soil health.

The undisturbed soil structure in no-till systems fosters a thriving environment for beneficial soil organisms. Earthworms, beneficial fungi, and bacteria flourish in this habitat, creating a complex web of life that enhances nutrient availability. These organisms break down organic matter, releasing essential nutrients that plants can readily absorb. This natural process reduces or eliminates the need for synthetic fertilizers, contributing to a more environmentally friendly approach to farming. Moreover, the increased soil organic matter improves soil structure, resulting in better water infiltration and drainage, making the soil more resilient to drought conditions. Healthy soil, teeming with life, forms the foundation of a successful and sustainable agricultural system.

No-till farming is not without its challenges. Weed management can become more complex, requiring careful planning and the implementation of effective weed control strategies. The initial transition to no-till can also be challenging, especially for farmers accustomed to traditional tillage methods. Proper equipment and knowledge are crucial to successful implementation. Understanding the different techniques for planting seeds directly into the residue, as well as managing potential pest and disease issues, is critical. However, these hurdles can be overcome with adequate training, planning and the use of appropriate technology.

The long-term benefits of no-till farming significantly outweigh these initial challenges. Reduced erosion safeguards the long-term fertility of the land, preserving its productivity for generations to come. The improved soil health and water retention capabilities enhance crop yields, often exceeding those of conventionally tilled fields over the long term, even in challenging conditions. Furthermore, reduced reliance on synthetic fertilizers and pesticides lowers the environmental footprint of the farm, reducing the pollution of water resources and air. The resulting savings in input costs further enhance the economic viability of no-till farming.

Implementing no-till farming necessitates careful planning and a gradual transition. It's vital to understand the specific soil conditions and climate of your region before making a full switch. Begin by experimenting with no-till techniques on a small scale, allowing you to learn the practical aspects and fine-tune your approach before applying it to larger areas. Seek guidance from experienced no-till farmers and agricultural experts. Attend workshops, join online forums, and connect with organizations dedicated to promoting sustainable agriculture. This will provide the knowledge and support needed to navigate the transition effectively. Remember, embracing no-till is a journey, not a destination; continuous learning and adaptation are essential for long-term success.

The economic advantages of no-till farming are substantial. While initial investment in specialized equipment might seem daunting, the long-term savings on fuel, labor, and inputs such as fertilizers and pesticides quickly offset these costs. The improved yields, coupled with reduced input expenses, often lead to increased profitability. Moreover, the enhanced soil health and environmental benefits can make the farm more attractive to consumers, potentially allowing for the development of premium products and access to specialized markets, further boosting income potential. This resilience to changing environmental conditions reduces the risks associated with unpredictable weather patterns, enhancing long-term economic stability.

By embracing no-till farming, farmers actively contribute to the creation of a healthier planet. The reduced use of synthetic inputs minimizes pollution, protecting water resources and air quality. The increased carbon sequestration improves soil health and reduces greenhouse gas emissions. The enhanced biodiversity fosters a more resilient and thriving ecosystem. No-till farming offers a pathway towards creating sustainable agricultural practices which benefit both the environment and the agricultural economy. It represents a fundamental shift towards a more environmentally conscious and economically viable approach to food production. It's an investment in the future of agriculture and the well-being of our planet.

The future of sustainable farming rests, in part, on the widespread adoption of this crucial practice.

2.2 Cover Cropping

Cover cropping isn't just about throwing seeds on the ground; it's a strategic maneuver in the ongoing battle for soil health. Think of it as a dynamic, living mulch, a proactive shield protecting your soil from the elements and enriching it simultaneously. It's about building resilience into your land, preparing it for the next planting season, and creating a vibrant ecosystem teeming with life.

Selecting the right cover crop is key. The ideal choice depends heavily on your specific climate, soil type, and the main crop you intend to plant later. For example, legumes like clover or alfalfa excel at fixing nitrogen in the soil, a crucial nutrient often depleted by intensive farming. Grasses like rye or oats excel at suppressing weeds and preventing erosion. Choosing a diverse mix of species can offer a broader range of benefits, strengthening your soil's natural defenses against pests and diseases. Remember to always consult local agricultural resources or extension services; their knowledge of regionally suitable species can be invaluable.

Planting time significantly impacts the success of your cover crop. Ideally, you'll sow your seeds after your cash crop is harvested, leaving ample time for the cover crop to establish itself before winter. However, you might also consider intercropping certain fast-growing cover crops with your primary crop to maximize benefits throughout the growing season. This technique requires careful planning and understanding of the specific growth rates of both plants to avoid competition for resources.

Properly managing your cover crop involves more than just planting. You need to monitor its growth regularly, ensuring that it's thriving and not becoming overly competitive with your main crop should you choose to interplant them. This monitoring allows for timely intervention, should issues such as pest infestation or disease outbreak arise. Early detection significantly improves your chances of successful mitigation, potentially preventing larger, more costly problems later. A well-maintained cover crop is a happy cover crop. Termination is equally important. Before planting your next cash crop, you'll need to terminate your cover crop. This could involve mowing, rolling, or tilling—the best approach again depends on your specific cover crops, soil type and climate. However, minimizing soil disturbance is usually advantageous, promoting a healthy soil structure that enhances water infiltration and prevents erosion. Leaving the cover crop residues on the soil surface to decompose naturally adds organic matter and improves soil fertility, setting the stage for a vigorous next crop.

Cover cropping offers multiple advantages extending far beyond immediate soil health improvements. By suppressing weeds, it reduces the need for herbicides, a significant win for environmental stewardship. The improved soil structure leads to better water infiltration and retention, reducing the need for irrigation in many instances and also bolstering drought resistance for your main crop. Moreover, cover crops often provide habitat for beneficial insects, contributing to the overall biodiversity of your agricultural ecosystem.

Integrating cover cropping into your farming practices represents a shift in mindset – it is a long-term investment, not just a quick fix. The initial costs might seem higher, but the long-term gains in soil health, reduced reliance on chemical inputs, and increased crop yields generally outweigh any initial expenses. Remember to document your results, meticulously tracking your progress over several growing seasons. This data will prove invaluable for refining your cover cropping strategy, adjusting your techniques, and demonstrating the cost-effectiveness and environmental benefits of your approach.

While the initial learning curve may seem steep, the wealth of resources available makes it a manageable endeavor. Your local extension office, agricultural universities, and various online forums offer a wealth of information, technical assistance, and even the opportunity to connect with experienced cover cropping practitioners. Don't hesitate to reach out and tap into these invaluable networks. Remember to tailor your approach to your unique conditions and learn from both your successes and challenges. Through patient observation and continuous adaptation, cover cropping can become an integral, highly rewarding component of your farming system.

The transition to cover cropping may require adjustments to your existing farming practices. For example, you might need to allocate additional time for seed selection, planting, and management. You may also find that your equipment requires minor adjustments to handle the different cropping systems. However, these minor inconveniences are far outweighed by the considerable long-term benefits. Consider cover cropping as an investment in the future fertility and productivity of your land.

Above all, patience and persistence are crucial. You won't see immediate, dramatic changes, but the cumulative effects over time are remarkable. Every successful cover crop planting contributes to building soil structure, improving water retention, and enriching the soil with vital nutrients. The more you work with cover crops, the more intuitive and effective your approach will become. Your land will thank you for it, season after season. And you will find a more sustainable, more resilient, and ultimately more successful farming system. The payoff is a healthier environment, healthier crops, and a healthier bottom line in the long run. Embrace the process, and you'll soon appreciate the transformative power of cover cropping.

2.3 Crop Rotation

Crop rotation, a cornerstone of sustainable agriculture, is far more than simply switching crops each year. It's a strategic dance between plant families, a carefully orchestrated ballet designed to improve soil health, boost yields, and minimize pest and disease pressures. Think of it as a long-term investment in your land, a commitment to its vitality that yields substantial returns over time.

Unlike monoculture farming, which depletes the soil of specific nutrients and leaves it vulnerable to specific pests and diseases, crop rotation diversifies the soil's nutritional needs. Legumes, for example, with their nitrogen-fixing capabilities, enrich the soil, leaving a legacy of fertility for the following crop. This natural fertilization minimizes the need for synthetic fertilizers, reducing both costs and the environmental impact associated with their production and use.

The benefits extend beyond nutrient management. Different crops have varying root systems, impacting soil structure. Deep-rooted crops like alfalfa break up compacted soil, improving aeration and water infiltration, creating a more hospitable environment for subsequent crops with shallower roots. This enhanced soil structure also improves water retention, crucial in combating the effects of drought. Consider how the deep taproots of legumes improve drainage, while shallow-rooted plants help prevent erosion and moisture loss.

Pest and disease control is significantly enhanced through crop rotation. By strategically alternating crops, you disrupt the life cycles of many common pests and pathogens. A pest that thrives on one crop might find itself without a suitable host in the following season, significantly reducing its population and limiting the damage it can inflict. This natural pest management minimizes reliance on chemical pesticides, contributing to safer food and healthier ecosystems. The economic advantages of this are clear, reducing expenditure on chemical treatments and improving the marketability of your produce.

The selection of crops for rotation is crucial and should be tailored to your specific conditions, soil type, and market demands. A well-planned rotation should consider the nutrient demands of each crop, its growth habit, and its susceptibility to pests and diseases. Local expertise, agricultural extension services, and research on regionally suitable crop rotations can greatly assist in this planning process. For instance, rotating between cereals, legumes, and root crops offers a balanced approach, addressing nutrient cycling, soil structure, and pest control effectively.

Moreover, crop rotation benefits extend beyond the farm gate. The reduced reliance on chemical inputs translates to cleaner water, healthier ecosystems, and a diminished carbon footprint. This aligns perfectly with the growing consumer demand for sustainable and environmentally responsible agriculture, leading to improved market access and potentially higher prices for your produce. Customers are increasingly willing to pay a premium for goods produced using sustainable practices. This translates into a more profitable and resilient agricultural enterprise. Therefore, thoughtful planning offers considerable long-term benefits.

Understanding the intricacies of crop rotation requires dedicated learning and ongoing observation. It's not a one-size-fits-all solution, and successful implementation demands careful consideration of your specific circumstances. Keeping detailed records of your crop rotations, noting yields, pest occurrences, and soil analyses, will allow for continuous improvement and refinement of your approach over time. This ongoing data collection creates a powerful tool for decision-making, adapting your strategies to changes in weather patterns, market demands, or pest pressures.

The transition to crop rotation might seem daunting initially, but the benefits far outweigh the challenges. Starting small, experimenting with a few strategically chosen crops on a portion of your land, will allow you to gain experience and confidence before expanding your implementation. Sharing experiences and knowledge with fellow farmers can also prove invaluable, fostering a community of shared learning and support. Remember, crop rotation is a journey of continuous learning and adaptation, a testament to the enduring commitment to sustainable agricultural practices. You are not simply growing crops; you're nurturing the very foundation of your agricultural success: healthy soil. And this investment will reap rewards for years to come, ensuring a prosperous and sustainable agricultural future. The positive effects on soil health, environmental sustainability, and economic viability will be clear in both the short term and the long term.

In conclusion, adopting crop rotation practices is a fundamental shift towards a more sustainable and economically sound approach to agriculture. It's not merely a technique; it's a philosophy, a commitment to the long-term health and productivity of your land. By understanding its intricacies and carefully implementing a well-planned rotation, you are not only improving the yields of your crops but also investing in a future where agriculture and environmental stewardship thrive in harmony. This will improve your overall success and help you sustain a profitable agricultural enterprise. Through ongoing observation, meticulous record keeping, and community engagement, you are ensuring that your efforts in crop rotation contribute to building a sustainable and thriving agricultural future, one harvest at a time. It is a key element in the creation of a truly resilient and prosperous agricultural system.

3 Chapter 3: Water Management

Water is lifeblood for agriculture; its efficient use is paramount. Ignoring this fundamental truth leads to reduced yields and environmental damage. We must move beyond traditional, wasteful methods. Consider drip irrigation, a technique that delivers water directly to plant roots, minimizing evaporation and runoff. This targeted approach significantly reduces water consumption compared to flood irrigation, a technique where water is applied across the entire field, often leading to significant losses. Sophisticated systems even monitor soil moisture levels, ensuring plants receive only the necessary amount. These systems are initially more expensive, but the long-term savings in water and energy, plus the improved yields, often offset the initial investment. Furthermore, the reduced water usage minimizes the environmental impact, promoting more sustainable agricultural practices.

Efficient irrigation is not simply a matter of technology; it requires careful planning. Understanding your soil type is crucial; sandy soils drain quickly, necessitating more frequent but smaller irrigations, while clay soils retain water longer, allowing for less frequent watering. Analyzing your specific climate is equally important. A hot, dry climate demands different irrigation strategies compared to a cooler, more humid one. Consider installing weather monitoring systems to adapt irrigation schedules according to rainfall and evapotranspiration rates, optimizing your water use in accordance to local conditions. The careful integration of technology, soil understanding, and climate awareness optimizes water efficiency.

Beyond efficient irrigation lies water harvesting, a proactive approach to conserving this precious resource. This involves collecting rainwater through various methods, from simple rain barrels to elaborate catchment systems. Rainwater harvesting reduces reliance on groundwater, thus preventing its depletion, a significant concern in many regions. Moreover, harvested rainwater is naturally free of harsh chemicals often present in treated water, making it ideal for irrigation and overall healthier for plants. The implementation of such strategies differs based on the scale of farming operation, local climate, and available resources. Larger operations may incorporate extensive reservoir systems, while smaller farms can use simple techniques like contour farming or terracing to manage water flow. The accumulated water can be stored in reservoirs or tanks, providing a reliable water source during dry spells.

Implementing water harvesting necessitates a holistic approach. Consider the landscape's topography when designing the collection system. Slope, soil type, and the surrounding vegetation all influence the efficiency of water collection and storage. Proper maintenance is essential too. Regular cleaning of gutters and reservoirs prevents sediment buildup and contamination, maintaining the quality of the harvested water. It is an investment, but the long-term benefits far outweigh the upfront costs. Remember, water scarcity poses a significant threat to global food security; effective water management is not merely beneficial but necessary for long-term agricultural viability. Integrating efficient irrigation and water harvesting techniques represents a substantial step toward building more resilient and sustainable agricultural practices. The long-term vision of sustainable agriculture necessitates such responsible stew-

ardship of resources.
Furthermore, water management isn't solely about technology or infrastructure; it demands a shift in mindset. Farmers must become astute water managers, learning to assess their needs, and adjusting their practices according to the available water resources. This includes selecting drought-tolerant crops and employing water-efficient farming methods. Accurate soil moisture monitoring allows farmers to irrigate only when necessary. Regular training and workshops can equip farmers with the needed skills and knowledge. Such proactive management ensures that water use is optimized, minimizing waste and maximizing yields even in water-scarce environments. This approach fosters better adaptation to climate change effects, and building more secure and sustainable farming systems. This conscious approach contributes significantly to the long-term sustainability of agricultural practices.
Finally, collaboration and community engagement are integral to effective water management. Sharing knowledge and best practices within farming communities promotes a collective approach to water conservation. Government agencies and research institutions play a vital role in providing support, disseminating information, and fostering collaborative efforts. Government policies and subsidies that incentivize water-efficient farming techniques play a significant part. By working together, farmers and governmental bodies can overcome the challenges posed by water scarcity and create a more resilient and water-secure future for agriculture. This collaborative framework ensures the widespread adoption of sustainable practices, contributing to long-term agricultural sustainability, and promoting efficient and responsible resource management. The future of agriculture hinges on such collective action and informed choices.

3.1 Efficient Irrigation

Efficient irrigation is crucial for maximizing crop yields while minimizing water waste, a vital concern given the increasing scarcity of this precious resource. Consider drip irrigation, a method delivering water directly to plant roots through a network of tubes and emitters. This targeted approach minimizes evaporation and runoff, significantly reducing water consumption compared to traditional flood or furrow irrigation. Implementing drip irrigation requires an initial investment in materials and installation, but the long-term savings in water and increased crop yields often outweigh the upfront costs. Moreover, carefully selecting appropriate emitters based on your specific crops and soil type is essential for optimal performance.

Beyond the choice of irrigation method, efficient water management involves understanding your soil's water-holding capacity. Conducting a soil test helps determine the appropriate irrigation frequency and amount. Over-watering can lead to root rot and nutrient leaching, while under-watering results in stunted growth and reduced yields. Regular monitoring of soil moisture levels, using either soil probes or moisture sensors, allows for precise irrigation scheduling, preventing both extremes. This data-driven approach, combined with weather forecasting, allows you to adjust your irrigation schedule proactively, optimizing water use and plant health.

Scheduling irrigation strategically is paramount. Irrigate deeply and less frequently rather than shallowly and often. Deep watering encourages deeper root growth, making plants more resilient to drought conditions. This also reduces the frequency of irrigation events, saving time and labor. Furthermore, consider the time of day when irrigating. Early morning irrigation minimizes evaporation losses, as the sun's intensity is lower. Avoid irrigating during the hottest part of the day, as much of the water will be lost to evaporation before reaching the plant roots.

Optimizing irrigation efficiency also extends to maintaining your irrigation system. Regularly check for leaks, blockages, or damaged emitters. Promptly address any issues to prevent water loss and ensure even water distribution. Cleaning your irrigation system periodically, especially in areas prone to mineral buildup, maintains its efficiency and extends its lifespan. A well-maintained system ensures that water is delivered precisely where it's needed, preventing wasteful runoff or uneven watering.

Water harvesting, a technique often overlooked, plays a crucial role in supplementing your irrigation needs. Collecting rainwater in tanks or reservoirs provides a valuable source of water, especially in regions with limited rainfall. This collected water can be used to irrigate crops during dry periods, reducing reliance on external water sources. Implementing water harvesting involves installing appropriate collection systems, depending on your scale of operation and the availability of space. Proper storage and filtration of harvested water are essential to prevent contamination and maintain water quality.

Beyond these practical methods, consider the broader implications of water use in agriculture. Adopting drought-resistant crop varieties reduces your reliance on irrigation, conserving water resources. Utilizing mulching techniques helps retain soil moisture, reducing the need for frequent watering. Furthermore, educating yourself on water-efficient agricultural practices and participating in community-based initiatives promotes collective water conservation. Small changes across a larger community contribute significantly to better water management and sustainability.

Choosing the right irrigation method, coupled with careful monitoring and maintenance, drastically improves water use efficiency in your agricultural practices. It's about more than just getting water to your crops—it's about managing that resource wisely for both your yields and the environment's health. By understanding soil conditions and implementing smarter irrigation scheduling, you'll not only ensure thriving crops but also contribute towards responsible water stewardship. This is paramount in agriculture's ongoing quest for sustainable practices, ensuring future food security amidst climate change. Remember, the journey to efficient irrigation involves continuous learning, adaptation, and a commitment to responsible resource management. The rewards, however, are abundant—healthier crops, reduced water bills, and a smaller ecological footprint.

3.2 Water Harvesting

Water harvesting is more than just collecting rainwater; it's a proactive strategy for enhancing agricultural resilience, especially in areas with unpredictable rainfall. Think of it as creating a safety net for your crops, ensuring they have access to moisture even during dry spells. This proactive approach minimizes the reliance on increasingly scarce and expensive irrigation resources. Implementing effective water harvesting systems requires careful planning and consideration of your specific environment and needs.

Successful water harvesting begins with assessing your land's topography. Identify areas where water naturally collects, like low-lying spots or depressions. These are prime locations for constructing your harvesting system. The size of your system will depend on factors such as the size of your farm, average rainfall, and the water needs of your crops. Remember, a small, well-managed system is better than a large, poorly designed one. Consider consulting a local expert to help you make informed decisions about system size and location.

Constructing a water harvesting system involves creating structures to capture and store water. This could involve building simple

swales or terraces to direct runoff towards designated collection points, or more complex systems like reservoirs or ponds. Materials used in construction should be locally sourced whenever possible to reduce your environmental impact and support local economies. Durable and appropriate materials, such as concrete, clay, or even repurposed materials, are crucial for system longevity. Construction should be done in a way that minimizes soil erosion and protects the surrounding environment. Simple, low-tech solutions are often the most practical and effective, especially for small-scale farmers.
The choice of water storage method significantly influences your system's overall efficiency. Simple earthen reservoirs, while inexpensive, may require more maintenance to prevent seepage and evaporation. Covered tanks and lined reservoirs are more effective at retaining water but may require a larger upfront investment. Regardless of your choice, proper sealing and maintenance are essential to prevent water loss and contamination. Regular cleaning and inspections will ensure the longevity and efficiency of your storage method.
Beyond the physical structures, effective water harvesting requires a strong understanding of water management techniques. This includes optimizing your water use, minimizing evaporation, and implementing strategies to protect your stored water from contamination. Techniques like mulching and drip irrigation can help reduce evaporation from the soil. Protecting your reservoir from livestock and other sources of contamination is crucial for maintaining water quality. Regular water testing can help you identify and address any potential issues early on.
The integration of water harvesting with other sustainable agricultural practices is key to maximizing its benefits. For example, combining water harvesting with crop rotation and cover cropping can help improve soil health, increase water retention, and enhance the overall efficiency of your farming system. These integrated ap-

proaches create a synergistic effect, leading to greater resilience and productivity. By combining these practices, you're creating a holistic approach to sustainable farming, moving beyond simple water storage to a comprehensive resource management strategy.

Understanding the legal and regulatory framework surrounding water harvesting is crucial to avoid potential conflicts and ensure compliance. In some regions, there may be specific regulations governing the construction and use of water harvesting systems. Researching local laws and obtaining any necessary permits before embarking on your project is essential. Local authorities can provide guidance and support to ensure you're complying with regulations. Proactive engagement with your local community can also help to avoid potential disputes and foster cooperation.

Finally, consider sharing your knowledge and experience with other farmers. Water harvesting is a practice that benefits everyone. Participating in workshops or training programs can further enhance your understanding and contribute to broader community adoption. By sharing your knowledge, you are contributing to the creation of a more resilient and sustainable agricultural community. Through collective learning and sharing, we can ensure a future where water is managed sustainably and agricultural productivity is enhanced for years to come.

4 Chapter 4: Biodiversity's Impact

Healthy, diverse ecosystems are the cornerstone of successful and sustainable agriculture. Think of it like this: a vibrant, bustling city needs a diverse population to thrive; similarly, a thriving farm needs a diverse range of plants, insects, and other organisms working together. This chapter will explore this vital connection, focusing on practical steps farmers can take to harness the power of biodiversity. Ignoring this crucial element is like building a house on sand—it might look sturdy at first, but it's inherently unstable.

Consider the humble honeybee. These industrious creatures are responsible for pollinating a significant portion of our food crops. Their decline poses a direct threat to our food security, underscoring the critical role of pollinators in maintaining agricultural productivity. Supporting pollinator populations, by planting diverse flowering plants around fields and reducing pesticide use, is an act of safeguarding our food supply. This is not just a matter of environmental concern, it's a direct investment in the future of food production. These efforts provide a crucial safety net against future pollination shortages.

Beyond bees, numerous beneficial insects contribute significantly to pest control. Ladybugs, for example, are voracious aphid eaters, naturally reducing the need for chemical pesticides. Lacewings and praying mantises are also powerful allies in the fight against crop-damaging insects. Encouraging these beneficial insects through habitat creation—leaving areas of wildflowers and avoiding broad-spectrum insecticides—is a far more sustainable and environmentally friendly approach than relying on potentially harmful chemicals. This strategy, while potentially demanding more upfront planning and effort, delivers a healthier, more environmentally sound result in the long run.

Natural enemies play a crucial role in regulating pest populations, preventing outbreaks that can devastate crops. These predators and parasites maintain a natural balance, keeping pest numbers in check without the need for extensive intervention. Understanding these natural relationships allows farmers to work with the ecosystem rather than against it, creating a self-regulating system that is both efficient and environmentally responsible. This harmonious approach reduces the farmer's reliance on external inputs, promoting economic and environmental independence.

The benefits of biodiversity extend far beyond pest control. Diverse ecosystems offer numerous ecosystem services crucial for agricultural productivity. Healthy soil, rich in organic matter and teeming with life, is a direct result of biodiversity. Microbial communities break down organic matter, releasing nutrients that feed plants, enhancing their growth and resilience. This improved soil health reduces the need for artificial fertilizers, conserving resources and reducing the environmental impact of agriculture. The benefits ripple through the entire agricultural process.

Furthermore, biodiversity enhances resilience to climate change. Diverse cropping systems are more likely to withstand extreme weather events such as droughts and floods. They are also more resistant to pests and diseases. This resilience translates into stable and reliable crop yields, mitigating the risks associated with climate change—a growing concern for farmers worldwide. These diversified systems act as a buffer against the unpredictability of the climate, giving farmers greater confidence and stability.

Incorporating biodiversity into agricultural practices is not just environmentally responsible; it's also economically advantageous. Reduced reliance on chemical inputs translates into lower costs, higher profit margins, and ultimately, a more sustainable agricultural business model. Consumers are increasingly demanding sustainably produced food, making biodiversity-friendly farming a valuable marketing asset. This shift in consumer demand presents an excellent opportunity for farmers to differentiate their products and command higher prices.

So, how can farmers practically integrate biodiversity into their operations? Creating habitat diversity is paramount. This can involve planting hedgerows, leaving field margins un-cultivated, and incorporating diverse cover crops into rotations. Minimizing pesticide use is also critical, allowing beneficial insects and other organisms to thrive. Employing integrated pest management strategies, which incorporate biological controls alongside other methods, is a more sustainable and effective approach to pest management. This holistic approach promotes a more harmonious co-existence between crops and the surrounding ecosystem.

Ultimately, harnessing biodiversity's power is not merely an optional add-on to sustainable agriculture; it is a fundamental requirement for long-term success. It is a holistic approach that integrates ecology and economics, creating resilient, productive, and environmentally friendly farms. Embracing biodiversity is an investment in the future—a future of healthy food systems, thriving ecosystems, and profitable, sustainable farms. By working with nature, not against it, farmers can create a truly sustainable and successful agricultural enterprise. The benefits are substantial and far-reaching, impacting not only the farmer's bottom line, but also the health of the planet and the well-being of future generations.

4.1 Pollinator Power

Bees, butterflies, moths, even bats – these aren't just pretty creatures; they're the unsung heroes of our food supply. Their tireless work pollinating crops is essential for a vast majority of the fruits, vegetables, and nuts we enjoy. Without them, our plates would look dramatically different, and our economies would feel the impact. Think of the vibrant colors of a strawberry patch, the sweet scent of an apple orchard, the buzzing hive – these are all testaments to the power of pollination. This chapter isn't just about the cute creatures though, it's about the practical steps you can take to support them and, in turn, support the future of agriculture.

Protecting these pollinators is more than just an environmental concern; it's a matter of economic security for farmers. A healthy pollinator population directly translates to higher crop yields and improved quality. Conversely, declining pollinator numbers mean lower yields, higher costs, and increased food insecurity. The economic implications are far-reaching, affecting not only farmers but also consumers and the entire food system. We are talking billions of dollars in potential losses, a stark reminder of the critical role these creatures play.

One simple yet effective way to enhance your farm's attractiveness to pollinators is by diversifying your crops. Instead of focusing solely on monoculture, consider incorporating a variety of plants. Include flowering plants with different bloom times to ensure a consistent food source for pollinators throughout the growing season. This creates a more resilient and sustainable ecosystem, beneficial for both your crops and the pollinators. This also boosts biodiversity on your land.

Avoid using harmful pesticides, especially broad-spectrum insecticides that indiscriminately kill beneficial insects alongside pests. Opt for targeted pest control methods, such as biological control or integrated pest management (IPM) strategies that minimize harm to non-target organisms. Remember, pollinators are vital parts of a healthy ecosystem. Protecting them requires careful consideration of the chemicals used on your land. Research into safer alternatives is crucial for the long-term health of both your crops and the environment.

Creating nesting habitats is another crucial step. Many pollinators, such as bees, need safe places to nest and raise their young. This can be as simple as leaving patches of undisturbed ground or installing bee hotels. Planting native flowering plants also provides a natural habitat for native pollinators. Supporting local biodiversity is fundamental for long-term pollination success. Remember, diverse habitats means diverse pollinators, strengthening the overall pollination power.

Educate yourself and others about the importance of pollinators. Spread the word about the threats they face and the simple actions individuals can take to help. This awareness campaign might involve local community groups, workshops, or school programs. The more people understand the significance of pollinators, the more likely they are to support conservation efforts. Raising awareness builds a stronger, more supportive community around the agricultural sector.

It's also crucial to advocate for policies that support pollinator conservation. Support initiatives that promote sustainable agricultural practices, protect natural habitats, and regulate the use of pesticides. Your voice matters – engage with your local representatives and support organizations working to protect pollinators. This proactive step can influence larger-scale changes that are beneficial for agriculture and the environment as a whole. Don't underestimate the power of your voice.

By actively supporting pollinator populations, you're not only ensuring the success of your own farm but also contributing to a more sustainable and resilient food system for generations to come. Remember, we have a collective responsibility to protect these vital creatures and their invaluable contribution to our food security. Protecting them is protecting our future. The long-term health of our food systems is directly linked to the health and abundance of pollinators. The work begins with simple, yet profoundly impactful, actions.

4.2 Beneficial Insects

Lacewings, with their delicate, gossamer wings, are voracious aphid predators. Their larvae are particularly effective, consuming hundreds of aphids during their development. Introducing lacewings to your farm can significantly reduce aphid populations, minimizing the need for chemical pesticides and promoting a healthier ecosystem. Consider purchasing lacewing eggs or larvae from a reputable supplier, releasing them strategically throughout your fields, focusing on areas with high aphid infestations. Regular monitoring will help you assess their effectiveness and make adjustments as needed. Remember, a thriving lacewing population depends on a diverse habitat with ample nectar and pollen sources for adult feeding.

Ladybugs, those cheerful red and black beetles, are another powerful ally in your fight against pests. Both adult ladybugs and their larvae are voracious feeders of aphids, mealybugs, and other small insects that can damage your crops. Attract ladybugs to your farm by planting flowers rich in pollen and nectar, such as dill, fennel, and yarrow. These flowers will provide sustenance for the ladybugs, encouraging them to stay and lay their eggs. Avoid using broad-spectrum insecticides, as these can inadvertently harm ladybugs and other beneficial insects. A healthy ladybug population is a natural and effective pest control method.

Ground beetles are often overlooked but play a crucial role in soil health and pest control. These nocturnal predators scurry through the soil, hunting for slugs, snails, caterpillars, and other soil-dwelling pests. They are especially effective at controlling pests that may not be visible on the plant itself. Maintaining a healthy soil structure with adequate organic matter is key to attracting and supporting a robust ground beetle population. Avoid practices that disrupt the soil ecosystem, such as excessive tillage, and focus on methods that promote biodiversity within the soil. Their presence indicates a healthy and balanced environment.

Praying mantises, with their distinctive raptorial forelegs, are formidable predators. These insects will hunt down and devour a wide range of pests, including caterpillars, grasshoppers, crickets, and even other insects that might damage your crops. They're incredibly efficient hunters and can significantly reduce pest populations. Providing suitable habitat for them is important – tall grasses, shrubs, and other vegetation will attract them to your farm, providing them with places to hunt and lay their eggs. These are effective bio-control agents you'll want to encourage.

Syrphid flies, often mistaken for bees or wasps, are beneficial insects that act as pollinators and also consume significant numbers of aphids. Their larvae are especially effective aphid predators, consuming large numbers of these pests. Adult syrphid flies are attracted to flowers, so planting a variety of flowering plants will encourage them to visit your fields. They contribute to pollination, increasing crop yields while reducing the need for chemical pest control, offering a two-fold benefit. This makes them valuable assets in your integrated pest management strategy.

Hoverflies, similar to syrphid flies, are excellent pollinators and their larvae feed on aphids. These are beneficial insects that can be a substantial part of an integrated pest management plan. Adult hoverflies feed on nectar and pollen from flowers, while their larvae are effective aphid predators. By providing a diverse range of flowering plants, you can attract and support a healthy hoverfly population. This contributes to both pollination and pest control, showing the interconnectedness of agricultural ecology. A balanced ecosystem benefits from these.

Trichogramma wasps, tiny parasitic wasps, are highly effective in controlling various lepidopteran pests such as caterpillars. The female wasps lay their eggs inside the eggs of their host insects. The wasp larvae then develop within the host egg, preventing the pest from hatching. These are microscopic champions in the fight against crop-damaging insects. Commercial releases of Trichogramma wasps are a common practice in some agricultural settings. Timing is critical; releases must coincide with the peak egg-laying period of the target pest. Careful planning and monitoring are essential for effective biological control through these insects.

These are only a few of many beneficial insects. Remember that diverse habitats are essential to attract and support a wide range of beneficial insects and other organisms that help create a naturally balanced and productive farm. By understanding the roles of beneficial insects, farmers can employ a variety of strategies to encourage their presence, decreasing dependence on chemical pest control and creating a more sustainable and resilient agricultural system. This natural approach improves soil quality, crop yields, and overall ecosystem health.

4.3 Pest Control

Integrated pest management (IPM) is your best friend. Forget the blanket spraying of harsh chemicals; it's damaging to the environment and ultimately less effective in the long run. IPM focuses on prevention and monitoring, using a combination of methods to keep pest populations below damaging levels. Think of it as a holistic approach, understanding the entire ecosystem within your farm. It

prioritizes natural solutions, minimizing harmful interventions.
Regular scouting is key. Walking your fields, inspecting your crops for signs of pests, and noticing changes in their behavior – this vigilance is crucial. Early detection allows for timely intervention before infestations become widespread. Remember, a small problem is much easier to solve than a large-scale one, saving you resources and stress. Don't just look at the plants themselves, but also the surrounding environment. Observe the soil, nearby plants, and even the air. Look for signs of unusual activity or changes in the ecosystem. Understanding the early warning signs allows you to adopt a preventative approach, rather than having to react to a major infestation.

Consider companion planting. Certain plants have natural pest-repelling properties. Marigolds, for example, are known to deter nematodes and other soil-borne pests. Basil can help repel many insects that attack tomatoes. Strategic planting can create a naturally pest-resistant environment, reducing the need for external interventions. Research which plants work well together in your specific environment and crop rotation. It's like creating a natural barrier, protecting your crops from pests.

Beneficial insects are your allies, not your enemies. Ladybugs, lacewings, and praying mantises are voracious predators of many common agricultural pests. Encourage their presence by providing habitat and food sources. Avoid broad-spectrum insecticides, as these can harm beneficial insects alongside the pests. Protecting these natural predators offers a powerful, eco-friendly method of pest control, allowing them to naturally regulate populations. Encouraging biodiversity is fundamental to this approach, promoting a thriving ecosystem.

Cultural practices play a significant role in pest control. Proper crop rotation disrupts pest life cycles, making it difficult for them to establish themselves. Healthy soil, rich in organic matter, supports strong, resilient plants better able to withstand pest attacks. Maintaining good soil health creates a strong foundation for plant growth and pest resistance. It is a crucial component of integrated pest management, ensuring a balanced ecosystem that minimizes the need for external control methods.

Physical barriers can sometimes work wonders. Row covers can protect seedlings from pests, and sticky traps can effectively monitor pest populations and catch a substantial number of insects. These methods offer direct and targeted action, limiting environmental impact and ensuring that the method is specific and efficient. This targeted approach helps to reduce the need for broad-spectrum insecticides.

When chemical intervention is absolutely necessary, select the least-toxic option and apply it judiciously. Always follow label instructions carefully and consider the impact on beneficial insects and the environment. Even if you have to use chemicals, opt for targeted and specific approaches. This ensures that the control is effective and precise while minimizing environmental damage. Remember, you are only using this as a last resort, having already employed the previous methods of pest control. Always consider the long-term environmental impact.

Biological control agents can be effective in managing specific pests. These are naturally occurring organisms that prey on or parasitize pests. Introducing them into your farming system can be a sustainable and effective alternative to chemical control. Carefully consider your options based on your specific pest and local conditions; you need to make sure this biological control agent will actually work and won't negatively impact other organisms. Remember, this approach requires thorough research and careful planning to be effective.

Document your efforts. Keep a detailed record of your pest monitoring, control methods, and their effectiveness. This information will allow you to refine your strategy over time and make more informed decisions in the future. This ensures that you're adapting your pest control strategy to your environment and the changing conditions. Continuous learning and documentation is key to success in the field. Through the continuous adjustment of your methods, you will find the best strategy for your specific environment.

4.4 Natural Enemies

Harnessing the power of natural enemies is a cornerstone of integrated pest management (IPM), a strategy that minimizes reliance on synthetic pesticides. Instead of eradicating all insects, IPM focuses on maintaining a balanced ecosystem where beneficial insects, acting as natural predators or parasites, keep pest populations in check. This approach requires a keen understanding of the intricate web of life within your fields and the specific organisms that thrive in your particular environment. Remember, what works exceptionally well for one farmer might not work for another, given the vast diversity across agricultural settings.

Effective IPM necessitates careful observation. Spend time regularly inspecting your crops, noting the presence of both pests and beneficial insects. Identify the pests; understanding their life cycle is crucial for effective control. Then, identify the beneficial insects. Lacewings, ladybugs, and parasitic wasps are just a few examples of the many natural enemies that can significantly reduce pest populations. These beneficial insects are often overlooked, yet they are invaluable allies in the fight for healthy crops.

Knowing the specific natural enemies of your target pests is critical for successful implementation of your IPM plan. For example, if you're dealing with aphids, introducing ladybugs or lacewings into your field can prove remarkably effective. Similarly, parasitic wasps can be highly effective against certain moth larvae. Researching the specific natural enemies prevalent in your region and suitable for your crops is a crucial first step. This requires a blend of practical knowledge and targeted research. Local agricultural extension offices and universities often provide valuable resources and expert advice.

Introducing natural enemies to your farm is more nuanced than simply releasing them. Success hinges on providing a suitable habitat. This involves creating diverse habitats within and around your fields to support the life cycles of your beneficial insects. This might include planting flowering plants to attract pollinators, leaving undisturbed areas of vegetation for shelter and overwintering, and avoiding the use of broad-spectrum pesticides that could inadvertently harm beneficial insects alongside the pests. Maintaining healthy soil is also crucial; a thriving soil ecosystem supports a robust population of beneficial organisms.

Beyond introducing new organisms, consider enhancing the habitat to attract existing natural enemies. Avoid practices that harm beneficial insects. Minimize or eliminate the use of broad-spectrum insecticides; these indiscriminate chemicals often wipe out beneficial insects along with the pests they control. Instead, consider using targeted treatments against specific pests, employing them only when absolutely necessary and after carefully assessing the situation. This requires patience and observation, letting natural processes do their work whenever possible.

Implementing an IPM program involves a commitment to careful monitoring and adjusting your approach as needed. Regularly inspect your crops for signs of pest pressure and beneficial insect activity. This ongoing assessment lets you intervene effectively when pest populations begin to rise above acceptable levels, but with precision and a minimization of disruption to the overall balance of the ecosystem. You can use visual inspections or more advanced techniques like pheromone traps, allowing for early detection of pest activity. By combining careful observation with data-driven decision-making, you can fine-tune your IPM program to be both effective and sustainable.

The transition to IPM may require adjustments in your farming practices. It can be a learning process and might not yield immediate results, requiring patience and perseverance. However, the long-term benefits are significant, including reduced pesticide costs, a healthier environment, and a more resilient cropping system. Remember to document your observations and results, so you can continually refine your strategies over time, making adjustments based on what works best in your unique circumstances. It's a journey of learning, adapting, and fine-tuning, but ultimately it's a path towards a healthier and more sustainable agricultural practice. The rewards—both ecological and economic—are well worth the effort. Successful IPM is far from a passive approach; it requires active management and ongoing adaptation. You must continually monitor pest and beneficial insect populations. The data you collect should guide your decisions, allowing you to fine-tune your strategy over time. Don't be afraid to experiment with different approaches and to learn from both successes and failures. It's a dynamic process, and flexibility is key to maintaining a healthy balance within your agricultural ecosystem. This is a collaborative process; seek advice from experienced farmers and agricultural professionals, leveraging the expertise available to optimize your approach. By embracing this adaptive and proactive approach, you'll unlock the full potential of natural enemies and cultivate a more resilient and sustainable farm.

4.5 Ecosystem Services

Healthy ecosystems are not just aesthetically pleasing; they're the bedrock of successful agriculture. These natural systems provide a wealth of services, often overlooked, that directly impact yields, reduce costs, and enhance the overall sustainability of farming practices. Understanding and actively supporting these services is paramount for long-term success in agriculture.

Consider pollination, a vital ecosystem service. Wild bees, butterflies, and other pollinators are far more efficient than human intervention, often resulting in superior fruit set and seed production. Their work is crucial for numerous crops, including fruits, vegetables, nuts, and even some forage crops. Protecting and enhancing pollinator habitats through practices like planting diverse flowering plants near fields becomes a crucial element of sustainable farming and should be planned out ahead. This strategy often results in higher yields and reduces the need for costly and environmentally damaging artificial pollination methods. Their presence is an indicator of a robust, healthy ecosystem within and surrounding the agricultural operation, leading to further benefits beyond pollination alone.

Moving beyond pollination, consider the role of beneficial insects in pest control. Ladybugs, lacewings, and parasitic wasps naturally prey on common agricultural pests. By fostering biodiversity, farmers create environments that support these natural predators, reducing the reliance on synthetic pesticides. This integrated pest management approach not only protects beneficial insects but also safeguards human health and minimizes environmental contamination. The cost savings from decreased pesticide use are significant, allowing a greater profit margin. Furthermore, consumers are increasingly showing a preference for pesticide-free produce, creating a profitable niche market for growers who implement these sustainable practices.

Beyond pest control, the soil itself benefits immensely from a healthy ecosystem. Mycorrhizal fungi, a type of beneficial soil fungus, forms symbiotic relationships with plant roots, enhancing nutrient uptake and water absorption. Earthworms and other soil organisms improve soil structure, aeration, and drainage, creating optimal conditions for plant growth. The soil's organic matter content, boosted by the activity of these organisms, improves water retention and reduces erosion. Maintaining a diverse and thriving soil ecosystem naturally provides these benefits, dramatically reducing the need for expensive fertilizers and soil amendments. The health of the soil and subsequent yield increases directly affect profit margins, highlighting the financial benefits of biodiversity.

Furthermore, healthy ecosystems contribute to climate regulation. Trees and other vegetation act as carbon sinks, absorbing atmospheric carbon dioxide, which is a vital factor in mitigating climate change. This is especially true for crops that incorporate cover crops, which are a fantastic method to increase organic material in the soil and sequester carbon. The reduction of carbon in the atmosphere and the improvement of soil health have a significant positive impact on the environment. This carbon sequestration is particularly relevant with our changing climate, which is rapidly becoming a defining factor in the agricultural success or failure of a region.

In essence, ecosystem services are not merely an added bonus for sustainable agriculture; they are fundamental components of a successful and resilient farming system. By understanding and actively promoting the health of the ecosystems that surround and underpin your operation, farmers can significantly improve their yields, reduce costs, enhance their environmental footprint, and create more resilient and profitable businesses. This is particularly important, as the climate is becoming increasingly erratic and unpredictable. Investing in ecosystem health is an investment in the future of your farm and the wider agricultural landscape. Protecting these services is not just environmentally responsible; it's financially smart. The long-term benefits far outweigh any perceived initial costs.

5 Chapter 5: Economic Sustainability

Farmers often struggle with fluctuating market prices, impacting their income and long-term viability. Fair pricing mechanisms, therefore, are crucial for economic sustainability. This necessitates strong farmer cooperatives and advocacy groups that negotiate better deals with buyers, ensuring producers receive a just return for their labor and produce. Supporting local farmers' markets and community-supported agriculture (CSA) programs can also help create more direct connections between farmers and consumers, fostering fairer pricing and strengthening community ties. A transparent and traceable supply chain, allowing consumers to understand the journey of their food, further empowers fair pricing initiatives by building consumer trust and willingness to pay a premium for ethically sourced products.

Market access is another critical component of economic sustainability in agriculture. Many small-scale farmers lack access to larger markets, limiting their sales potential. Improving infrastructure, such as better roads, transportation networks, and cold storage facilities, is essential to overcome these logistical hurdles. Government initiatives promoting access to export markets are particularly beneficial for farmers producing specialty crops or goods with high export potential. Online marketplaces and e-commerce platforms offer alternative avenues for market access, enabling farmers to reach a wider customer base, reducing reliance on intermediaries, and boosting profitability. However, ensuring digital literacy and technological access among farmers is crucial for successful implementation.

Value-added products offer significant opportunities for increased income and economic resilience. Transforming raw agricultural produce into processed goods, such as jams, juices, or dried fruits, can significantly increase the value and profitability of a farm's output. This requires investment in processing equipment and marketing expertise, but the resulting higher profit margins can greatly enhance farm sustainability. Diversification into value-added products also reduces reliance on single commodity prices, mitigating the risk associated with market fluctuations. Developing niche markets for unique or specialty products can further enhance profitability and attract premium prices, establishing a stronger market position for farmers. Training programs focusing on food processing techniques, business planning, and marketing strategies are essential to enable farmers to successfully venture into value-added product development.

Economic sustainability in agriculture necessitates a holistic approach that integrates fair pricing, improved market access, and the development of value-added products. It requires collaboration between farmers, policymakers, and consumers to create a supportive environment for profitable and sustainable agricultural practices. Financial literacy programs for farmers, helping them to manage their finances and plan for the future, are also crucial in building financial resilience. Investing in agricultural research and development helps develop higher-yielding and more resilient crops, enhancing productivity and overall farm profitability, supporting economic sustainability. Moreover, access to micro-loans and other forms of credit can provide farmers with the capital they need to invest in their businesses and expand their operations, directly contributing to their economic stability.

The creation of farmer support networks and cooperatives allows farmers to share resources, knowledge, and best practices, strengthening their collective bargaining power and fostering a supportive business environment. Education and training programs are vital in equipping farmers with the skills and knowledge needed to adapt to changing market conditions, adopt new technologies, and implement sustainable farming practices. Ultimately, economic sustainability in agriculture is about creating a system that is fair, equitable, and resilient, empowering farmers to thrive while protecting the environment and ensuring food security for generations to come. The journey towards economic sustainability requires continuous learning, adaptation, and a commitment to building strong, resilient farming communities. By focusing on these key strategies, we can pave the way for a future where agriculture is not only sustainable ecologically but also economically viable for all.

5.1 Fair Pricing

Fair pricing isn't just about covering costs; it's the cornerstone of a thriving agricultural community. Farmers, the backbone of our food system, deserve compensation that reflects their labor, investment, and the inherent value of their sustainably produced goods. Ignoring this fundamental principle jeopardizes the future of agriculture itself, leading to farm closures and a compromised food supply. A fair price ensures a sustainable livelihood, motivating farmers to continue their vital work and invest in improving their practices. Furthermore, it encourages the adoption of sustainable methods, as farmers are properly rewarded for their efforts in environmental stewardship.

Profit margins are often squeezed thin due to fluctuating market prices and escalating input costs. This instability directly impacts the farmer's ability to reinvest in their operations, hindering innovation and improvements that ultimately benefit consumers through better-quality products and more efficient production. To navigate these challenges, farmers need access to diversified markets and strategic partnerships that help buffer against price volatility. Developing local food systems and fostering direct consumer relationships can provide vital stability and fairer prices. Farmers' markets and community-supported agriculture (CSA) programs offer excellent examples of this approach.

Understanding the true cost of production is crucial. This involves far more than simply calculating expenses related to seeds, fertilizers, and labor. It should encompass the full environmental footprint, including soil health improvement, water conservation efforts, and biodiversity enhancement. These often go uncompensated, but their long-term value to both the environment and the community is immense. Transparency in pricing should reflect this wider range of contributions, recognizing and rewarding the farmer's dedication to sustainable practices. This could involve developing certification programs and labeling systems that highlight the environmental and social benefits of sustainably produced food.

Negotiating fair prices requires empowerment. Farmers need to collaborate, forming cooperatives and producer organizations to strengthen their bargaining power in the marketplace. Collective action allows for greater influence in negotiations with buyers and processors, ensuring fairer prices that better reflect the true value of their produce. Education and training are critical components of this empowerment, equipping farmers with the skills to manage their finances, understand market dynamics, and effectively advocate for their interests. Knowledge is power, and empowering farmers to understand and navigate the economic landscape is fundamental to fair pricing.

Beyond simple monetary value, a fair price recognizes the social and cultural importance of farming. Agriculture is deeply intertwined with our history and identity, nurturing communities and shaping landscapes. Farmers maintain rural vitality, preserving cultural heritage, and contribute to the economic resilience of their regions. Recognizing this wider context is essential to establishing a fair compensation model that truly values the farmer's multifaceted contributions. Supporting policies that prioritize the wellbeing of farming communities, such as land access initiatives and infrastructure development, are key components of this approach.

To further promote fair pricing, consumers also have a critical role to play. Being aware of where your food comes from, and choosing to support farmers who prioritize sustainability and fair labor practices, makes a substantial difference. Paying a slightly higher price for sustainably produced goods directly supports farmers in adopting and maintaining these crucial methods. Educating consumers about the benefits of sustainable agriculture and the importance of fair prices can create a virtuous cycle, with consumers making informed decisions that reward sustainable farmers and enhance the resilience of the food system as a whole. This conscious consumerism creates a market demand that effectively incentivizes fair practices throughout the agricultural supply chain.

Moreover, government policies and regulations can be instrumental in establishing a level playing field for farmers. Policies that support farmer cooperatives, regulate unfair market practices, and provide access to credit can all contribute to fairer pricing. Furthermore, investments in research and development that support the adoption of sustainable farming techniques can help reduce production costs and improve profitability. These measures, along with effective monitoring and enforcement mechanisms, are essential to creating a just and sustainable agricultural economy. Ultimately, the goal is to create a system that rewards environmentally responsible and

socially equitable agricultural practices.
In conclusion, achieving fair pricing for agricultural products is not merely an economic issue; it is a matter of social justice and environmental sustainability. It requires a multi-faceted approach involving farmers, consumers, and policymakers working collaboratively to build a more equitable and resilient food system. Fair pricing enables farmers to thrive, communities to flourish, and the environment to prosper, ultimately paving the way for a more sustainable and secure food future for generations to come. This is a journey that requires collective action, sustained commitment, and a shared understanding of the importance of fair pricing as a foundational element of a thriving agricultural ecosystem.

5.2 Market Access

Farmers often struggle to connect their produce to consumers directly, a challenge hindering profitability and economic stability. Establishing strong market access is crucial, requiring a multifaceted approach. It's not simply about getting your goods to market; it's about building relationships, understanding market demands, and finding the best channels to reach your target customers. Consider your unique selling points; what makes your farm and products stand out? Highlighting this individuality is critical in a crowded marketplace.
Direct-to-consumer sales, such as farmers' markets and Community Supported Agriculture (CSA) programs, provide excellent avenues for building direct relationships with customers. These channels allow you to control pricing, interact directly with your customers, build brand loyalty, and receive immediate feedback. However, the logistics involved, such as transportation and storage, must be carefully managed.
Consider partnering with local restaurants, grocery stores, or food

cooperatives. These collaborations offer broader reach than direct sales alone. Building positive relationships with buyers is paramount, emphasizing consistent quality and timely deliveries. Negotiate fair prices that reflect the value of your sustainable practices. Remember, you are selling not just produce but also a story of commitment to soil health and environmental stewardship. This narrative resonates deeply with consumers increasingly interested in ethically sourced, sustainable food.

Explore online marketplaces, leveraging the power of e-commerce. Building a user-friendly website or utilizing established online platforms opens access to a wider geographic area. High-quality product photography and clear descriptions are essential, capturing the attention of potential buyers. Effective digital marketing is key, utilizing social media and targeted advertising campaigns. Online sales require additional planning for packaging, shipping, and handling potential logistical complications.

Value-added products significantly increase profit margins and attract customers looking for convenience. Transforming your raw produce into jams, pickles, sauces, or other processed goods expands market possibilities and reduces reliance on selling only fresh produce. This strategy requires additional processing equipment and expertise, but the higher returns often justify the investment.

Community involvement plays a crucial role in strengthening market access. Participating in local events, sponsoring community initiatives, and fostering strong relationships with neighbors fosters trust and brand awareness. These actions increase visibility and create a loyal customer base. Consider offering workshops on sustainable gardening or cooking classes using your produce to actively engage the community.

Networking with other farmers, agricultural organizations, and industry professionals can provide valuable insights and collaborative opportunities. Sharing knowledge, resources, and market access

strategies helps build a supportive and thriving agricultural community. Participating in trade shows and conferences opens doors to new markets and partnerships.
Finally, remember that securing market access isn't a one-time event; it's an ongoing process requiring constant adaptation and improvement. Regularly assess your market performance, gather customer feedback, and adjust your strategies as needed to adapt to shifting demands and opportunities. The dynamic nature of the food industry requires flexibility and a willingness to continuously innovate. This constant adaptation is vital for long-term success.

5.3 Value-Added Products

Adding value to your farm's produce significantly boosts profitability and resilience. It's not just about selling raw goods; it's about transforming them into higher-priced, more desirable items. Consider the simple act of transforming fresh tomatoes into sun-dried tomatoes; the shelf life extends dramatically, opening up new markets and commanding higher prices. The increased value isn't just about the price; it's about meeting a consumer demand for convenience and unique flavors, driving sales even further.
Think about specializing in niche markets. Perhaps your region is known for a specific type of apple, but you could go beyond simply selling the fruit. Creating apple cider, apple butter, or even a unique apple-based sauce creates a whole new revenue stream, tapping into a different sector of the market with a higher profit margin. This diversification reduces your reliance on single crop sales and provides a safety net against market fluctuations. The key is identifying your unique regional strength and developing products that capitalize on it.
Farmers' markets and direct-to-consumer sales offer fantastic opportunities. You can showcase your value-added products directly

to customers, forging stronger relationships and receiving immediate feedback. This direct engagement allows you to build a loyal customer base and command premium prices, reflecting the care and craftsmanship put into your products. It's a direct line to the consumer, bypassing the middleman and maximizing your profit. Consider offering workshops or demonstrations showcasing your process—adding another dimension to your value proposition.

Strategic partnerships can significantly expand your reach. Collaborating with local chefs, restaurants, or food retailers can create a steady demand for your products. This isn't just about supply and demand; it's about building a local food system that benefits everyone. The collective marketing and increased visibility open doors to a much larger consumer base than you could reach independently, increasing your chances of success. Furthermore, exploring partnerships with artisans or other producers can lead to collaborative ventures, expanding your product line and brand reputation.

Processing your own produce, even on a small scale, opens up many possibilities. Making jams, jellies, pickles, or other preserved foods extends the shelf life of your crops and reduces waste. These value-added products can be sold at farmers' markets, directly to consumers, or through partnerships with local retailers. Such diversification reduces your reliance on immediate sales, allowing for a more stable income stream and providing a buffer against seasonal fluctuations in demand. Even small-scale processing can significantly impact profitability.

Investing in appropriate equipment and training is crucial for success. Whether it's a simple dehydrator for drying fruits and vegetables or more advanced equipment for processing dairy products, investing in quality tools is essential for efficiency and high-quality output. Investing in training ensures the safe and efficient use of this equipment, maximizing your return on investment while maintaining the highest quality standards. The cost of the equipment is an investment, not an expense, leading to higher profits in the long run.

Packaging and presentation matter tremendously. Attractive and informative packaging helps your products stand out on shelves and at farmers' markets. Consumers are drawn to products that look good, and clear labeling highlights your brand and product value. Invest in creative labels that tell a story and emphasize your commitment to sustainable practices. This attention to detail is integral to building your brand and communicating your commitment to quality.

Marketing your value-added products requires a comprehensive strategy. This includes creating a strong brand identity, utilizing social media platforms to connect with consumers, and participating in local events to increase brand awareness. You need a plan to communicate the unique selling points of your products and engage directly with potential customers. A well-crafted marketing strategy will ensure your value-added products reach their target market effectively.

Finally, remember that continuous learning and adaptation are essential. The world of food and agriculture is constantly evolving, so stay informed about new technologies, consumer trends, and market demands. Flexibility and willingness to experiment are vital to staying ahead of the curve and maintaining a profitable and sustainable business. Always seek feedback from consumers and adapt your strategies based on their responses. Continuous improvement is the key to long-term success.

6 Chapter 6: Technological Advancements

Precision farming, a cornerstone of modern agriculture, leverages technology to optimize resource utilization and enhance crop yields. Imagine a farmer meticulously monitoring soil conditions, nutrient levels, and even individual plant health, all from the comfort of their office, using data gathered from sensors strategically placed across their fields. This is the reality that GPS-guided machinery, remote sensing, and sophisticated data analytics bring to the forefront. Such precision allows for targeted application of fertilizers, pesticides, and irrigation, minimizing waste and maximizing efficiency. No longer are broad applications the norm; instead, each area receives exactly what it needs, reducing environmental impact and lowering input costs. This technology also allows for improved crop monitoring, enabling quicker identification and responses to potential problems.

Data analysis plays an increasingly crucial role in this technological revolution. The sheer volume of data generated by sensors and other monitoring devices necessitates advanced analytical capabilities. Sophisticated algorithms, leveraging machine learning and artificial intelligence, can analyze vast datasets to identify patterns, predict future yields, and optimize resource allocation. This predictive capability is transforming decision-making in agriculture, enabling proactive measures rather than reactive ones. For example, farmers can anticipate potential pest outbreaks or nutrient deficiencies, allowing for timely interventions and prevention of significant crop losses. Moreover, by integrating data from weather patterns, soil conditions, and market prices, farmers can make informed decisions regarding planting schedules, crop selection, and marketing strategies. This data-driven approach moves agriculture towards a more scientific and efficient model.

The integration of these technologies requires a significant shift in mindset and approach. Farmers need access to the necessary training and resources to effectively utilize these tools. This involves not only acquiring the equipment but also understanding the data interpretation and decision-making processes. Furthermore, access to reliable internet connectivity and sufficient digital literacy are critical for effective implementation and utilization of data-driven insights. The challenges are not just technical; equally crucial is the economic aspect. The initial investment in precision farming technology can be substantial, potentially creating a barrier for smaller farmers. Finding affordable and accessible financing options is therefore critical for wider adoption.

However, the long-term benefits far outweigh the initial investment. Reduced input costs, increased yields, and minimized environmental impact translate into enhanced profitability and sustainability. This creates a positive feedback loop, allowing farmers to reinvest profits into further technological upgrades and improvements. The integration of such technologies must be carefully considered to ensure equitable access and avoid exacerbating existing inequalities within the agricultural sector. Government policies and support programs play a pivotal role in making these advancements accessible to all farmers, fostering a more inclusive and sustainable agricultural landscape. The development of user-friendly interfaces and training programs is also essential for wider adoption, empow-

ering farmers to confidently harness the power of technology.
Moreover, the use of drones and remote sensing technologies provides farmers with an aerial perspective of their fields, offering unparalleled insights into crop health and growth patterns. High-resolution imagery allows for quick detection of disease outbreaks, nutrient deficiencies, or areas of water stress. This enables immediate targeted interventions, minimizing losses and preventing large-scale problems. The combination of real-time monitoring and advanced data analytics provides farmers with a comprehensive understanding of their fields, fostering a proactive and data-driven management approach. Beyond crop monitoring, drones can be instrumental in tasks such as targeted spraying of pesticides or fertilizers, further enhancing precision and minimizing environmental impact. This efficient use of resources translates into significant economic and environmental gains. The ongoing advancements in drone technology, such as extended flight times and improved sensor capabilities, continue to enhance its applicability across various aspects of agricultural operations.
In conclusion, the integration of technology into modern agriculture is not just about adopting new tools; it represents a paradigm shift in how we approach farming. Precision farming and data analysis, while demanding initial investment and training, ultimately offer enhanced efficiency, reduced costs, increased yields, and minimized environmental impact. However, the successful and equitable integration of these technologies necessitates thoughtful consideration of access, affordability, and training. A collaborative effort involving farmers, researchers, policymakers, and the private sector is essential to ensure that these advancements benefit all stakeholders and contribute towards a more sustainable and resilient agricultural future. The journey towards a technologically advanced and sustainable agriculture is a collective one, requiring continuous innovation, collaboration, and a commitment to responsible implementation.

6.1 Precision Farming

Precision farming hinges on gathering and analyzing data to optimize resource use. Imagine a farmer, not haphazardly scattering seeds across a vast field, but instead precisely planting each one where it will thrive most, based on soil analysis, moisture levels, and even the specific needs of each plant. This level of control minimizes waste and maximizes yields, making it far more efficient than traditional methods. We're talking about a significant reduction in inputs like fertilizers and pesticides, meaning less environmental impact and healthier ecosystems.

This detailed approach relies heavily on technology. Sensors embedded in the soil monitor conditions in real-time, transmitting crucial data wirelessly to a central system. This constant stream of information—soil moisture, nutrient levels, temperature, even pest infestations—allows for immediate adjustments to irrigation, fertilization, and pest control strategies. GPS-guided machinery, meanwhile, ensures precise application of inputs, preventing overlaps and maximizing efficiency. Think of it as giving each plant exactly what it needs, when it needs it. It's targeted care, leading to healthier crops and a reduced ecological footprint.

Furthermore, the use of drones allows for aerial surveys, providing a bird's-eye view of the entire field. This facilitates early detection of crop stress, disease, or pest infestations, allowing farmers to intervene promptly and prevent larger problems from developing. Such proactive management reduces losses and minimizes the need for extensive interventions later on. The imagery provided is incredibly valuable, allowing for informed decision-making and enhancing overall farm management.

Data analysis is the key to unlocking the full potential of this technology. Sophisticated software programs analyze the immense amount of data collected, generating insightful reports that pinpoint areas needing attention. This allows farmers to adjust their strategies in real-time, maximizing yield and minimizing environmental damage. For instance, it might identify areas with nutrient deficiencies, allowing for targeted fertilizer application, preventing unnecessary chemical use elsewhere. The result is a sustainable system that delivers high yield with fewer resources.

Beyond just increasing yields, precision farming offers significant economic benefits. Reduced input costs—through efficient use of water, fertilizers, and pesticides—contribute directly to increased profit margins. Early detection of problems also reduces crop losses, adding further value. Moreover, many precision farming technologies are scalable, meaning they can adapt to farms of all sizes, from small family-run operations to large commercial farms. The accessibility of this technology makes sustainable agriculture achievable for many different actors in the industry.

However, precision farming isn't without its challenges. The initial investment in technology can be substantial, potentially presenting a barrier to entry for some farmers. Furthermore, the complexity of data analysis requires specialized skills and knowledge, creating a need for training and support programs to ensure farmers are equipped to fully leverage the technologies at their disposal. Access to reliable internet connectivity is also crucial, which can be a limiting factor in some rural areas, hindering adoption.

Despite these hurdles, the benefits of precision farming are undeniable. The enhanced efficiency, reduced environmental impact, and increased profitability make it a critical component of a sustainable agricultural future. It allows us to not just feed a growing global population, but to do so in a way that conserves our natural resources and promotes environmental health. As technology continues to advance, and as access to training and support increases, precision farming is set to become even more integral to agricultural practices worldwide. The future of farming is precise, and it's sustainable.

Ultimately, the success of precision farming depends on the integration of technology with sound agricultural principles and responsible stewardship. It's not just about technology; it's about using it wisely and ethically to achieve a balance between productivity and environmental responsibility. This approach necessitates a fundamental shift in mindset, moving from a reactive to a proactive approach to farming, maximizing yields while minimizing environmental impact. The rewards are not only financial but also ecological and social, building a more resilient and sustainable food system for generations to come.

Precision farming presents a transformative opportunity for agriculture, allowing farmers to become more efficient and environmentally conscious. This transition, however, requires careful planning and considerable investment in both technology and training. It's an evolution that necessitates collaboration between farmers, researchers, technology providers, and policymakers, ensuring that the technology's potential is fully realized in a manner that is both equitable and sustainable. The future of farming is not just about technology; it's about human ingenuity and responsible stewardship working in harmony with nature.

6.2 Data Analysis

Harnessing the power of data analysis in agriculture isn't just about collecting numbers; it's about transforming raw information into actionable insights that boost yields, optimize resource use, and enhance overall farm profitability. Think of it as a high-tech magnifying glass, allowing you to see patterns and trends invisible to the naked eye. This allows for more efficient decision making and reduced waste.

Effective data analysis begins with identifying your key performance indicators, or KPIs. What are the most critical aspects of your farm's operation? Is it maximizing yield per acre, minimizing water usage, or optimizing fertilizer application? Defining these KPIs early on ensures that your data collection efforts are focused and relevant, avoiding the trap of gathering unnecessary information. Prioritize what truly matters to your specific operations. The data should directly influence your decision-making.

Once you've established your KPIs, the next step is data collection. Modern agriculture offers a wealth of data sources, from yield monitors on your machinery to soil sensors measuring moisture levels and weather stations providing real-time climatic data. Smart irrigation systems, GPS-guided machinery, and even simple record-keeping apps contribute to a comprehensive dataset. The key is to integrate these different sources seamlessly, creating a holistic picture of your farm's performance. You must have a well-defined system that combines all the information you collect.

Data visualization is crucial. Simply collecting data is not enough. Raw data, even in large quantities, is meaningless without interpretation. Powerful software applications and dashboards allow you to present this data in a clear, digestible manner. Graphs, charts, and maps can highlight trends, outliers, and areas for improvement. Imagine seeing your field's yield mapped out, with clear areas of high and low productivity. This visual representation instantly pinpoints areas needing attention, facilitating targeted interventions. The clarity of the data presentation is vital to effective decision making.

Now, let's talk about specific applications. Precision farming relies heavily on data analysis. By analyzing soil samples, yield maps, and weather patterns, you can tailor fertilizer and irrigation applications to the specific needs of each section of your land. This minimizes waste, reduces environmental impact, and maximizes yields. Consider variable-rate technology, where fertilizers and pesticides are applied in varying amounts depending on the specific requirements of each area. This ensures efficient resource management.

Furthermore, predictive modeling allows you to anticipate challenges before they emerge. By analyzing historical data on weather, pests, and disease outbreaks, you can develop models to forecast potential problems. This enables proactive mitigation strategies, such as adjusting planting dates or implementing preventative measures. Early warning systems are becoming increasingly crucial in modern agriculture, helping to minimize financial loss and environmental damage. These models require accurate data sets and advanced analytical skills.

Beyond maximizing yield, data analysis offers cost-saving opportunities. Analyzing fuel consumption, equipment maintenance logs, and labor costs can reveal inefficiencies and areas for cost reduction. Consider using your data to track and improve the efficiency of your equipment. Identifying less-productive machinery could allow for a reduction in overall cost.

Remember that data analysis is not a one-time event; it's an ongoing process. Regularly review your data, update your models, and adapt your strategies as needed. This ongoing monitoring ensures that your farm's operations remain efficient and profitable. The ability to adapt and change is a core tenet of data analysis's usefulness.

Finally, consider seeking professional help if needed. Many agricultural consultants specialize in data analysis and can provide valuable insights and support. Their expertise can help you navigate the complexities of data management and interpretation. Don't hesitate to seek outside assistance; this can be beneficial to enhance your capabilities. This isn't just about technology; it's about making informed decisions based on accurate, up-to-date information that leads to better yields, reduced costs, and a more sustainable farm.

7 Chapter 7: Policy and Advocacy

Effective policy is the bedrock of sustainable agriculture. Without supportive legislation and robust funding mechanisms, even the most innovative farming practices will struggle to thrive. Government incentives for adopting regenerative techniques, such as tax breaks for implementing no-till farming or subsidies for cover cropping, can significantly accelerate the transition to more sustainable methods. These financial aids directly alleviate the initial financial burden faced by farmers, encouraging wider adoption. Furthermore, stronger regulations limiting the use of harmful pesticides and fertilizers can protect both the environment and farmer's health, fostering a healthier agricultural landscape.

Strong farmer support programs are essential. This goes beyond mere financial assistance; it involves access to resources, expertise, and mentorship. We need readily available training programs focusing on sustainable practices, equipping farmers with the knowledge to make informed decisions. These programs must go beyond lectures and offer hands-on experience, fostering a practical understanding of techniques. Establishing networks connecting experienced sustainable farmers with newcomers creates invaluable opportunities for learning and knowledge sharing, paving the way for collaboration and mutual support. Governmental and non-governmental organizations can actively promote these networks and facilitate these connections.

Education and training initiatives are not only crucial for farmers but also for the wider public. Educating consumers about the importance of sustainable agriculture increases demand for sustainably produced food, driving market changes. Integrating sustainable agriculture principles into school curricula familiarizes younger generations with these essential concepts, fostering future support for environmentally conscious practices. Additionally, accessible online resources, workshops, and community events can disseminate essential knowledge and promote widespread understanding and adoption.

Community engagement is vital in driving the adoption of sustainable agricultural practices. This requires creating platforms for dialogue between farmers, consumers, researchers, and policymakers. Local farmers' markets showcase the benefits of sustainable agriculture directly to consumers, fostering direct relationships and increasing the demand for environmentally friendly products. Community-based agricultural projects, involving local communities in initiatives promoting sustainability, build ownership and increase participation. Moreover, community-led initiatives can effectively advocate for policy changes that benefit their local agricultural communities, creating a powerful force for change.

Global collaboration is crucial because sustainable agriculture is a global challenge. Sharing best practices, research findings, and policy innovations across international borders accelerates progress. International organizations play a pivotal role in facilitating this cooperation, offering platforms for information exchange and technical assistance. Collaborating on research projects concerning climate-resilient farming techniques enables a coordinated approach to adapt to the impacts of climate change on food security. These international partnerships also strengthen agricultural systems, ensuring food security globally. Furthermore, international agreements and regulations help standardize sustainable agricultural practices, offering a framework for consistent implementation on a global scale.

7.1 Sustainable Policies

Sustainable agricultural policies must prioritize soil health. Protecting and enhancing soil fertility is fundamental; without healthy soil, there's no foundation for sustainable food production. Policies should incentivize practices like no-till farming and cover cropping, directly supporting farmers who adopt these regenerative techniques. Funding for research into soil health, especially tailored to specific regional conditions, is equally crucial. This isn't just about scientific advancement; it's about equipping farmers with the knowledge and tools to thrive. Furthermore, strong policies must enforce responsible land management, preventing soil erosion and degradation, and ensuring the long-term viability of agricultural lands.

Water conservation is another keystone of successful sustainable policies. Subsidies for efficient irrigation systems and rainwater harvesting technologies can significantly reduce water waste in agriculture. Moreover, policies should encourage the adoption of drought-resistant crops and water-wise farming practices, particularly in regions prone to water scarcity. These strategies contribute to both environmental protection and economic resilience, ensuring farmers can maintain production even in challenging conditions. The implementation and enforcement of such policies require substantial investment in infrastructure and education, ensuring a widespread understanding and adoption of water conservation strategies. Governmental support and partnerships with agricultural organizations are essential.

Policies supporting biodiversity are equally critical. This involves establishing protected areas within agricultural landscapes, promoting habitat diversity, and encouraging the use of integrated pest management strategies that minimize reliance on harmful pesticides. Supporting pollinator populations through the creation of pollinator-friendly habitats should be a priority, ensuring the continued pollination of crops. Furthermore, these policies should include incentives for farmers who actively integrate biodiversity into their farming practices. Financial rewards for biodiversity conservation efforts can transform conservation from a cost to a potential source of income for farmers. The long-term health of the agricultural ecosystem and the financial stability of farmers depend on a thriving biodiversity.

Economic sustainability for farmers requires a multi-pronged approach. Fair pricing policies, guaranteeing farmers receive a just return for their produce, are paramount. This necessitates market regulation to prevent exploitation and ensure competitive pricing. Simultaneously, policies facilitating access to markets are crucial, enabling farmers to reach wider consumer bases and secure better prices for their goods. Furthermore, government support for developing value-added products, such as processed foods or specialized agricultural items, can significantly increase farmer income and economic resilience. These strategies necessitate collaboration between government agencies, farmers' organizations, and processing facilities to create efficient and equitable systems.

Technological advancements must be incorporated sustainably. Policies should promote the adoption of precision farming technologies, enabling resource-efficient agricultural practices. Investing in data analysis tools helps farmers make informed decisions, improving yields while minimizing environmental impacts. However, access to these advanced technologies must be equitable, with programs specifically aimed at supporting smaller farms and farmers in marginalized communities. Therefore, support for technological adoption shouldn't exacerbate existing inequalities. Rather, it should empower all farmers to integrate technology effectively. Government-funded training programs and collaborations with technology providers are needed to bridge the technological gap and foster sustainable innovation.

Effective policy implementation needs strong advocacy and community engagement. Governmental support for farmer education and training initiatives is essential to promote the adoption of sustainable practices. Community engagement initiatives, fostering dialogue and collaboration among farmers, policymakers, and other stakeholders, play a vital role in shaping effective agricultural policies. Furthermore, international collaboration ensures the sharing of best practices and knowledge among nations, fostering global sustainability in agriculture. Such collaborations can help develop international standards and policies supporting environmentally and economically viable agricultural practices across the world.

Ultimately, sustainable agricultural policies must be dynamic and adaptable. Regular evaluation and adjustments based on scientific findings, farmer feedback, and environmental changes are critical for ensuring their long-term success and relevance. Policies should also consider the social and cultural contexts of agriculture, reflecting the needs and concerns of different farming communities. A holistic approach that integrates ecological, economic, and social considerations is needed to ensure a viable and resilient agricultural future. The commitment to continuous improvement and adaptation is essential for achieving true sustainability in agriculture, enabling future generations to cultivate a thriving and prosperous world.

7.2 Farmer Support

Farmers often face numerous challenges, from fluctuating market prices to unpredictable weather patterns. Their resilience and dedication are crucial to feeding the world, yet they are frequently underserved. Effective support systems are paramount, ensuring their livelihoods and the sustainability of agriculture. We need to move beyond mere charity and embrace proactive, comprehensive strategies that empower farmers to thrive.

Direct financial assistance can be a lifesaver, especially during times of crisis. Subsidies for seeds, fertilizers, or equipment can lessen the financial burden, helping farmers invest in improved practices and technologies. However, these programs should be carefully designed to avoid dependency and to promote responsible resource management, not encourage unsustainable farming practices. Grants and low-interest loans can offer similar benefits, providing crucial capital for improvements, innovation, and expansion. The key is strategic allocation, ensuring funds reach those who truly need them, and that the aid helps to advance sustainable farming practices.

Beyond financial aid, access to vital information is equally critical. Agricultural extension services, providing technical guidance and educational resources, are invaluable. These programs, however, must be tailored to the specific needs of diverse farming communities, considering local contexts and challenges. Farmers need practical, readily applicable knowledge, not just theoretical information. This means local demonstrations, hands-on workshops, and easy-to-understand guides on everything from soil testing to pest management. Regular communication and ongoing support are vital to ensure that farmers can effectively integrate this knowledge into their daily work.

Market access is another significant hurdle. Many farmers lack adequate connections to buyers, leading to unfair pricing and limited income. Government initiatives can help establish farmers' markets, cooperatives, or direct-to-consumer marketing strategies. These initiatives need to go beyond simply creating structures and should include the training and support needed for farmers to manage supply chains, pricing, and the market dynamics successfully. Promoting fair trade practices and advocating for policies that guarantee fair prices are also crucial. Farmers need assurance of a consistent market, allowing them to plan for the future and make long-term investments with confidence.

Investing in infrastructure is an equally essential aspect of farmer support. Reliable transportation networks are crucial to get produce to market efficiently. Improved irrigation systems and access to clean water sources can boost productivity and resilience to drought. Investing in rural infrastructure, such as roads, electricity, and communication networks, is a long-term investment that pays dividends in terms of economic growth and farmer wellbeing. It fosters a supportive environment for sustainable agriculture, allowing farmers to fully realize their potential. Improving access to essential services like healthcare and education in rural areas is just as important, enhancing the quality of life for farming communities.

Strong community networks provide crucial social support and knowledge sharing among farmers. Support groups, sharing best practices and coping mechanisms, can be immensely valuable. Government policies can encourage this sort of community building. Facilitating access to networking events, providing technical assistance for community-based projects, and supporting the development of cooperative ventures will help strengthen social support systems and build resilience within the farming community.

Finally, fostering innovation through research and development is essential to promoting sustainable and profitable agriculture. Supporting agricultural research, especially efforts focused on climate-resilient crops and improved farming techniques, can make a significant difference. Encouraging collaboration between researchers, farmers, and policymakers will help ensure that research results are translated into practical applications that benefit farmers directly. This collaborative approach is crucial for advancing sustainable practices and empowering farmers to meet the challenges of the future. Through a multi-faceted approach, we can ensure a thriving future for farmers and a sustainable food system for everyone.

7.3 Education & Training

Investing in the education and training of farmers is paramount for the success of sustainable agricultural practices. Without a skilled workforce, even the most innovative techniques and supportive policies will fall short. This necessitates a multi-pronged approach encompassing various learning styles and technological advancements to reach a wider audience and cater to diverse learning preferences. Farmers need access to up-to-date information on best practices, efficient technologies, and market trends. This knowledge empowers them to adapt to evolving challenges, improve their yields, and enhance the resilience of their operations.

Effective training programs should be tailored to the specific needs of the farming community they serve. This includes considering factors like the age, literacy levels, and prior experience of the participants. A diverse range of methodologies is crucial; from hands-on workshops and field demonstrations to online courses and interactive simulations. This ensures a comprehensive learning experience that caters to diverse learning styles. Consider pairing experienced farmers with newer generations, fostering a mentorship system that promotes knowledge sharing and collaboration within the community.

The curriculum should cover not only technical aspects, but also crucial business skills. Farmers need to understand financial management, marketing strategies, and risk assessment. They also need skills in negotiating fair prices with buyers and developing robust business plans to ensure the financial viability of their farms. Furthermore, integrating sustainable business models and responsible sourcing strategies are critical for building resilient agricultural enterprises. These models must also account for fair labor practices and environmental protection.

Technological literacy is equally important. Farmers need to be proficient in using precision farming tools, data analysis software, and other technological advancements. Training should emphasize the practical application of these tools, enabling farmers to make informed decisions based on data-driven insights, optimizing resource use and maximizing productivity. This also includes digital literacy – allowing access to online resources, market information, and communication networks.

Access to education and training shouldn't be limited to formal programs. Government agencies and non-governmental organizations (NGOs) play a critical role in facilitating access to resources. This includes organizing workshops, providing scholarships, and creating online learning platforms. This support system is crucial to overcome potential financial barriers to education and training. Collaborations with agricultural universities and research institutions are necessary to ensure the accuracy and relevance of the training materials.

The success of education and training initiatives depends on regular assessment and evaluation. Feedback mechanisms should be in place to identify areas for improvement. This may involve surveys, focus groups, or monitoring participant progress. Tracking the outcomes and impact of the programs will help identify best practices and optimize the allocation of resources. Continuous improvement is crucial to make these initiatives relevant and effective.

Governmental support is an essential element. This includes funding for education and training programs, providing incentives for farmers to participate, and creating a supportive regulatory environment. Policies that promote sustainable agriculture, along with financial support for farmers investing in technology and education, directly bolster these efforts. Additionally, encouraging public-private partnerships can help bring in additional resources and expertise.

Finally, community involvement is pivotal. Farmers learn from each other, and shared knowledge within a network builds a sense of shared purpose and collective action. This can be achieved through farmer cooperatives, peer-to-peer learning, and regular community gatherings. By fostering a collaborative learning environment, communities can build capacity and resilience, effectively tackling the challenges of sustainable agriculture. Building these strong community bonds ensures long-term success.

7.4 Community Engagement

Building strong community engagement is crucial for the success of any sustainable agricultural initiative. It's not merely about informing; it's about fostering a shared vision and collaborative action. This involves actively listening to diverse perspectives, understanding the unique challenges and opportunities faced by different community members, and creating platforms for open dialogue. Success hinges on recognizing that farmers aren't just producers; they are essential stakeholders whose knowledge and experience are invaluable.

Effective engagement necessitates creating a welcoming and inclusive environment, where everyone feels heard and respected. Avoid top-down approaches; instead, opt for participatory methods that encourage feedback and ownership. This might involve community forums, workshops, or even informal gatherings. The goal is to empower farmers and community members to become active participants, not passive recipients, of information. Remember, inclusivity means considering diverse groups within the community – women, youth, marginalized populations – ensuring their voices are amplified.

Consider implementing a multi-pronged approach encompassing diverse communication channels. Newsletters, social media platforms, local radio programs – all can serve as effective tools for disseminating information and fostering two-way communication. Ensure the information is accessible, tailored to different literacy levels and communication preferences. Using simple language, avoiding jargon, and employing visual aids can greatly enhance understanding and engagement.

Transparency and accountability are fundamental to building trust. Openly share information about project plans, budgets, and progress. Regularly solicit feedback, and demonstrate how this feedback is incorporated into decision-making processes. This proactive approach fosters trust and reduces the likelihood of misunderstandings or conflicts.

Beyond communication, think strategically about capacity building. Provide training opportunities on sustainable agricultural practices, market access, and financial management. Empowering farmers with the knowledge and skills they need to participate effectively is paramount. This also increases the community's resilience and self-sufficiency.

Remember the long game. Community engagement is not a one-off event but a continuous process requiring sustained effort and investment. Building strong relationships takes time. Regular communication, consistent support, and celebrating successes together are essential for building a thriving and resilient community committed to sustainable agriculture. The collaborative effort will inevitably

create strong bonds.
Financial sustainability is often overlooked in community engagement initiatives. Secure diverse funding sources, balancing government grants, private donations, and even potential revenue generation from community-based projects. This financial stability enables continuous engagement and ensures long-term viability.
Finally, recognize that conflict is inevitable, and disagreements are opportunities for growth. Establish clear mechanisms for resolving conflicts, fostering open dialogue, and finding solutions that are acceptable to all parties. Mediation or conflict resolution training for community leaders can be invaluable tools in promoting peaceful and productive collaborations. This approach fosters a healthy environment where constructive disagreement can lead to stronger outcomes. The key is to transform conflict into a constructive force for change, learning from differing viewpoints to achieve common goals. By fostering a culture of collaboration and mutual respect, the agricultural community will not only thrive but become a model for other sectors.

7.5 Global Collaboration

Global collaboration is paramount for achieving sustainable agriculture. It's not just about sharing best practices; it's about building a unified front against the challenges threatening food security worldwide. This requires a multifaceted approach, encompassing knowledge sharing, collaborative research, and the harmonization of policies across borders. Without a global commitment, individual efforts remain fragmented, hindering the potential for widespread positive change. The interconnectedness of our global food system demands a collective response.
Effective collaboration necessitates open communication channels. Farmers, researchers, policymakers, and consumers need to engage

in constructive dialogue, sharing experiences and knowledge across geographical boundaries. This can involve establishing international networks, facilitating knowledge exchange programs, and fostering partnerships between institutions and organizations working towards common goals. The free flow of information is crucial for rapid innovation and adaptation to changing environmental conditions. This open dialogue ensures diverse perspectives inform agricultural strategies.

One crucial aspect is collaborative research. Addressing complex issues like climate change and pest resistance demands a global research effort. Pooling resources and expertise allows for tackling large-scale challenges more effectively. International collaborations can facilitate the development of climate-resilient crops, improved farming techniques, and efficient water management strategies that are adaptable to diverse conditions around the world. Sharing the results through readily accessible and unbiased channels speeds progress.

Harmonizing agricultural policies across nations is another essential aspect. Inconsistent regulations can create trade barriers and hinder the adoption of sustainable practices. Working together to establish common standards and regulations promotes fair trade, supports environmentally conscious practices, and reduces negative impacts on global ecosystems. This also requires the equitable allocation of resources. We must ensure that developing nations aren't left behind.

Financial collaboration is critical for supporting sustainable agriculture in developing countries. Many lack the resources necessary to implement sustainable practices, access advanced technologies, or adapt to climate change. International development organizations, governments, and private sectors should collaborate in funding projects and programs supporting sustainable agriculture in these regions. This assistance needs to address the unique needs of each community, accounting for their specific environmental conditions and socio-economic contexts. Sustainable financing options are also crucial, including microloans, community investment funds, and fair trade initiatives.

Furthermore, technology transfer plays a significant role. Advanced technologies, such as precision farming techniques and data analytics, can significantly improve agricultural efficiency and sustainability. However, access to these technologies often varies considerably among nations. Global collaboration should focus on facilitating the transfer of appropriate technologies to developing countries, ensuring that these technologies are accessible, affordable, and adapted to local needs. This requires careful consideration of technological appropriateness and capacity building within the local communities.

Finally, raising global awareness about the importance of sustainable agriculture is crucial. Education and public awareness campaigns can promote environmentally conscious consumer choices and support for sustainable practices. Collaborative efforts involving governments, NGOs, educational institutions, and the media can help shape public perception and drive policy change. This broad public engagement is vital to ensuring sustainable agriculture becomes a shared global goal, not just a niche interest. The power of collective action cannot be overstated. Through collective efforts, we can nurture a more resilient, equitable, and prosperous future for all. The journey towards sustainable agriculture demands a global partnership, one which unites efforts and addresses common challenges. Each nation brings unique experiences and solutions to the table, creating a powerful collective resource for agricultural progress.

www.ingramcontent.com/pod-product-compliance
Lightning Source LLC
LaVergne TN
LVHW022024200926
848481LV00007B/2525